CAT BODY LANGUAGE

100 WAYS TO READ THEIR SIGNALS

CAT BODY LANGUAGE

100 WAYS TO READ THEIR SIGNALS

TREVOR WARNER

COLLINS & BROWN

First published in the United Kingdom in 2007 by
Salamander Books,
A division of the Pavilion Book Company Ltd.,
43 Great Ormond Street, London WC1N 3HZ

Distributed in the United States and Canada by
Sterling Publishing Co., Inc.
1166 Avenue of the Americas
New York, NY 10036, USA

All notations of errors or omissions should be addressed to Salamander Books,
43 Great Ormond Street, London WC1N 3HZ

ISBN: 978-1-911163-40-4

Printed in China

CONTENTS

INTRODUCTION

Twenty-first-century living is making the cat the most popular pet in the United States and Europe. Cats are not only fun to stroke, pleasing to look at, and great to have around, they have an independent streak that we admire. Though people treasure the trusting and faithful obedience of the dog, the high-maintenance pooch does not take itself out for a walk or throw itself sticks. In a world where leisure time is becoming increasingly limited, the self-contained, come-and-go-as-it-pleases cat is becoming king among pets.

Not only that, but more cats are becoming inside-only pets. Whereas a dog has too much raw energy for it to be kept inside, a cat can easily adapt to the confines of apartment living. The idea that Garfield of comic-strip fame is unrepresentative of the cat kingdom is soon disproved the minute you own one.

The huge boom in popularity of the Siamese, Burmese, and Tonkinese breeds is evidence that cat owners appreciate the companionship of cats, too. These are the most people-oriented of cat breeds, and owners often feel that their cats can understand their moods and emotions.

Cat behavior is governed by many things. Cats are not as sociable as dogs, but they can be sociable given the right circumstances—for instance, if they meet on what is regarded as neutral territory. Cats have the uneasy sociability of Western gunfighters coming into town. Whereas their dog compatriots can have a good woof at each other and indulge in canine posturing with very little risk if they give offense, cats have four sets of very sharp claws. With fragile ears to defend and potent weaponry on hand, the stakes are much higher when cats meet.

Cats prefer to deflect confrontation with suitable displays of threat and

appeasement. In this book, you will be able to see the telltale signs of cat aggression and submissive behavior courtesy of Jane Burton's remarkable photos. Cats have developed ways of communicating to other cats that they are not a threat in a bid to diffuse tension and avoid conflict, and some classic examples are seen on these pages.

Cats are territorial animals, mindful of their own ground and who comes onto it and when. Scenting plays a big part in their lives, and sniffing and marking boundaries is part of the daily routine of a cat. The cat that wants to be let out at night, and then twenty minutes later wants to come back in again is not indecisive. She's just out checking the latest scents.

When cats greet new people in their house with a rub of the head and a rub of the body, they are carefully leaving their scent marks on that person courtesy of scent glands on these areas. Humans, who are accustomed to touch relationships and not smell relationships, think that this contact comes from the cat because they are liked. However, cats will do the same to furniture and doorways, which are not considered objects of affection. Rubbing scent is a cat's way of putting its mark on the newcomer.

It's important for new owners to understand that cats live the lives of carnivores and the natural desire to kill is embedded in their genes. Like their big-cat cousins, they spend most of their time sleeping or lazing around, waiting for that hunting opportunity. When they eat, they pounce on their prey and devour a high-calorie meal that allows them to live a sedentary life the rest of the time.

The domestic cat has its meals provided, just as they were when they were kittens, so their life is one of extended kittenhood. Even though they have more than enough to eat, they will still hunt and eat what they catch. Cats cannot understand why they should be scolded for bringing home dead birds or half-chewed mice. They are providing this as a service to their "family" and do not expect to be reprimanded.

This book aims to explain all the facets of a cat's behavior: the way they act individually and also with other cats. Nowhere is an understanding of cat interaction more important than in a household with more than one cat, where the feline residents gently colonize certain areas for themselves and where a new introduction—such as a kitten or an adopted stray—can cause a huge upset in the hierarchy. In these situations, something as innocuous as moving the furniture around can have a profound effect.

This book will give you the first glimpse into the way a cat thinks. If you are still intrigued, there are many other books on the subject of cat psychology. However, if you are having problems with your cat's behavior, there is no substitute for taking your cat to see a trained animal psychologist for expert advice.

01 "I'M HAPPY AND CONTENTED."

Purring in cats, like tail wagging in dogs, is often misunderstood. A dog wagging its tail could still attack you because the tail wagging actually means "I'm excited," not "I'm pleased to see you." The purring cat is saying "I'm in an upbeat social mood." Though owners like to think it is their unadulterated love and careful stroking that causes the purring, cats will purr when they get their favorite food and even when they're injured or giving birth. It's also a signal intended to inspire a reaction from humans or other cats: "I want friendship," or "I want reassurance."

PURR-FECT HARMONY

Small cats can produce a satisfying purr, but so can some big cats. Lions, tigers, and cheetahs can do variations of the purr, but jaguars can't purr at all. The frequency of a cat's purr is anywhere between 25 and 150 hertz (Hz). At the bottom end of the range, around 25 Hz, the harmonics of a cat's purr are the same as an idling diesel engine.

02

"I WANT TO HEAR YOU ALL PURR."

Kittens begin to purr when they are about a week old and are suckling. It's a device that tells the mother all is well on the nipple and that each kitten is getting its share of milk. The mother will listen for their individual purrs and purr back to them as a reassurance that all is well, the nest is safe, and they can keep on suckling.

FREQUENT FLYER

Possibly the world's most traveled feline is a cat named Hamlet. He escaped from his carrier on a flight out of Toronto, Canada, and was given up as lost. He was found behind a panel in the aircraft seven weeks later, by which time he'd traveled 375,000 miles.

03 "I DON'T LIKE THIS. I SENSE DANGER."

Cats produce a caterwauling yowl in situations of extreme fearfulness. It's a low-pitched, mournful howl that varies in pitch. Cats yowl when they are about to confront another cat or execute an aggressive move. The yowl may start low and stay low as the two potential combatants give themselves a last-second chance to avoid a fight. As the pitch of the yowl rises, so the aggression commences. They can also yowl at vets, another situation of intense apprehension, locked as they are in their traveling carrier, surrounded by strange smells, and—if they have prior experience with the vet—anticipating pain.

THE SMALL CAT DRILL

Cats can be more expressive with their purrs than their big-cat cousins. Whereas a cat can create the purr noise as it breathes in and out—"like it's drilling behind the sofa," as comedian Eddie Izzard maintains—big cats such as lions and tigers can't. They can only produce the noise by moving air in a single direction—outward—no matter how contented they are.

04

"Aiiiiieeeee!"

At the moment of conflict, when the caterwauling ceases and fighting starts, cats will let out ear-splitting, high-pitched screams that last only a fraction of a second but wake up neighborhoods. The scream is a moment of sheer terror, as this kitten is experiencing when it turns and sees the dog it has been placed next to.

CAT LICK CHECKLIST

Cats have their own licking and cleaning checklist, and it's fascinating for owners to see if their cat performs it in the same way each time. The sequence is traditionally: Lick the paw and then rub it over one side of the head. Lick the other paw and rub it over the other side of the head. (Is your cat a lefty, a righty, or a don't-care-which-paw-first cat?) Lick shoulders, lick front legs, lick sides, lick genitals, lick hind legs, and lick tail.

05

"SORRY, MOM, I THOUGHT THIS TREE WAS EASY. NOW GET ME **OUTTA HERE!**"

The cat cry is used to summon help in an emergency. An example of a typical emergency is when a kitten goes too far in exploring its boundaries—in this case, getting stuck halfway up a tree without the courage to turn around and climb down. It's a pitiful, expressive cry that is designed to bring parental help.

06

"DON'T LOOK AT ME LIKE THAT!"

The hissing sound a cat makes is very similar to the hiss of a snake when it rears up, fearing attack. It's a sound reserved for close contact, when the aggressor can feel the full force of the air expelled over a cat's arched tongue. It has been theorized that cats developed this ability to imitate snakes through the evolutionary process. Dogs are extremely wary of snakes, and it is said that cats exploit this age-old fear. It's a nice theory, if it weren't for the fact that baby hedgehogs, when threatened, can also produce a loud, hissing, snorting noise made by expelling air rapidly. Cats may well be exploiting the surprise factor more than anything else.

# 07 "MEOW!"

It may surprise you to learn that cats rarely meow to each other. The meow is reserved almost exclusively for cat-to-human communication and is used to say "I'm here. Here I am. Look, I'm down here!" That is why there is often more meowing after a can of cat food is opened than beforehand, as the cat wants to make quite sure that it is the recipient. Meows are used primarily for food and access, and occasionally to say "hello," although some Burmese and Siamese owners will swear that they can have whole conversations with their cats.

USING UP ONE OF THE NINE

Florida senator Ken Myer thought he had lost his cat, Andy, when the cat fell from the sixteenth floor of his apartment building. Andy survived the fall, the longest recorded nonfatal fall by a cat. Also making use of its nine lives, a cat in Taiwan was discovered still alive after being trapped inside a collapsed building for eighty days following an earthquake in 1999.

08 "YOU CAN'T GO OUT LOOKING LIKE THAT."

In the extended kittenhood that is a house cat's life, stroking is a reminder that the owner is their "parent." Cats are one of the cleanest animals, and from the moment they are born, the mother will devote long hours to licking her litter clean. The abrasive cat tongue is a very good grooming tool, and kittens learn to be soothed and reassured by the action. Cat owners take over the role of "kitten mother" when they become responsible for feeding, protecting, and watching the kitten/cat. And just like Mom, the two-legged parent will also teach them proper grooming habits.

RECORD-BREAKING CAT

There are many claims for the world's oldest living cat. The problem for The Guinness Book of World Records is that most aged cats don't have verifiable records of their birth date. However, there are well-recorded instances of cats living into their thirties, and Creme Puff, a cat from Austin, Texas, celebrated her thirty-eighth birthday in 2005. A man in Dumfriesshire, Scotland, claimed to have a cat that was forty-three, but before the Scottish Cat Club could check out the details, it was killed by a train.

09

"HI, NEW MOM."

Stroking is a reassuring reminder of kittenhood and is enjoyed by even the oldest of cats. It is a maternal association that is deep in the cat's psyche. One thing a cat might do when it is first stroked on all fours is to hold its tail erect. This isn't a sign of excitement or a method of having the stroke extended along the back and then all the way up the tail. It is the action of a kitten who lifts his tail up to allow the mother to check that all is clean behind it. It is another sign of the cat reverting to kitten.

GOOD FUR YOU

Stroking a pet is widely viewed as a great way to relieve tension in humans, and cats have an advantage in that they have very soft and well-groomed coats. There are approximately 60,000 hairs per square inch on the back of a cat and about 120,000 per square inch on its belly.

10

"I FEEL SAFE WITH YOU!"

A cat lying around the house may see you approaching and roll over onto its back. This is the greeting of a (literally) laid-back cat. Whereas a more active cat might race up to you and start nudging its head against you, this cat is pleased to see you, but it's still pretty comfortable and warm where it is and doesn't want to trouble itself by getting up. The action of rolling over and exposing its vulnerable belly is something a cat won't risk if it senses any kind of danger. So it's an act of trust: "You're safe, but I still can't be bothered to get up."

"DON'T STROKE ME!"

Even though the rollover is a friendly, positive move, it is not an invitation to stroke the belly. The position is aimed at showing submissiveness. Some cats can be stroked here, but more often than not a stroke may be met with a swiped paw or even an attempted bite. This is a sensitive area, and a cat needs to be on very good terms with a human before the person is allowed to stroke its belly.

11 "HI, AND HERE'S MY SMELL."

A cat's greeting is very familiar. First, it will rub the top of its head against you, followed by the sides of its body, and with a final flourish, the tail. One pass is often not enough, and the friendly cat will come back with the other side of its body to leave a better impression. This initial contact from a cat often brings a response from cat-loving humans flattered by the attention, and they will reach down to stroke the cat—at which point the cat will rub its jaw into the stroking hand.

Though humans may not realize it, they are being marked with scent. There are scent glands on a cat's head, along its jaw, and at the base of the tail, and all these careful rubs are putting the scent of the cat on the guest. In return, the cat will smell the fur it has rubbed against us to figure out what we smell like. The fact that we don't bother to check the cat's scent is lost on the feline world. Cats like to make their presence felt throughout the house and will rub against baskets, furniture, and door frames. They are not saying hello to these objects.

12

"WHY DO YOU HAVE TO BE SO TALL?"

Cats would rather greet us nose to nose. It's the traditional cat greeting, after all. The fact that they can't is very frustrating to them, and they may occasionally get up on their hind legs to say hello when we come home. It's a vain attempt to get near our height and nuzzle. For this same reason, cats will often leap onto furniture to be closer to our height when we come through the door. They want to be able to say hello properly.

MAKING THE WRONG FRIENDS

While dog owners have more control over their animals, keeping them safe in the backyard and walking them on leashes, the wandering cat can make some very bad friends—especially a male who has not been neutered. The unspayed female will attract strange cats into the house when they are in heat or may try and escape to be "where the boys are," while unneutered tomcats will roam and bring home big veterinary bills after fights.

13 "I HOPE YOU DON'T MIND IF I . . ."

Cats will sniff each others' rear ends in greeting just like dogs do, except in the feline world it's a far more wary process. Cats will greet each other nose to nose first, and if they are friendly, one cat will allow the other to sniff its rear end. Then the process is repeated. They are not like dogs, who will charge up for a mutual sniff without any kind of preliminaries. It would be considered both a rude and aggressive move if a cat charged up for a sniff without any kind of engagement first. One of the reasons for this wariness is that cats have a lot more "armaments" ready to show disapproval—such as sharp claws. In this case, the tricolored cat has been far too quick in getting to know the ginger cat, even though his tail is raised. The direct stare from him is confrontational.

IS YOUR CAT A POLYDACTYL?

Though it sounds like a creature from Jurassic Park, a polydactyl cat is one with extra toes—sometimes six, sometimes as many as seven.

14 "HI THERE."

Cats like to rub noses and make nose-to-nose contact when they greet someone. So the simple solution for cat lovers who want to reinforce the bond between cat and human is to meet their cat nose-on. It can also form part of a scent greeting, in which scents are rubbed off and exchanged between cat and human. In this instance, the boy has picked up the cat so it can say hello, but cats are even happier if you come down to their level. That way, they have more control of what they do next and don't have to resort to "I want to get down now" actions to be let go.

SAY IT WITH MEOWS

Cats will often adopt a softer version of their meow as a "hello" meow. This muted version helps distinguish it from the louder "I want food" meow or the "let me out of the house" meow. Cats are far more sophisticated than dogs in expressing themselves vocally, and in some studies almost a hundred different vocal expressions have been recorded in cats, compared to under twenty in dogs.

"YOU SMELL FAMILIAR, AND I NEVER FORGET A NOSE."

When they meet in a neutral zone, cats that know each other say hello by sniffing and rubbing. First comes the nose-to-nose sniffing, which is followed by a little bit of head rubbing, perhaps some flank rubbing, and maybe some anal sniffing to finish it off. As described on page 35, this is usually done singly, with the more dominant cat being the first to sniff. It is also the cat who makes the first rubbing move that is the more dominant of the two. Here two kittens, who know each other very well, meet outside.

16 "NO SUCKLING TILL YOU'RE CLEAN."

Kittens are born at intervals of anything from a couple of minutes to an hour apart. A typical litter for a cat would be five or six kittens, so the process can be over quickly or last for several hours, by which time the mother will be exhausted. On average, kittens will arrive at twenty- to thirty-minute intervals, allowing the mother time to clean up each kitten and, most importantly, clear out the airways. Kittens are born inside an amniotic sac, which the mother will bite away before starting the cleanup phase. Here Alexandra welcomes her fifth kitten (of six) into the world by giving it a once-over lick.

LOOKING FOR CLUES

One big clue that a cat is about to give birth is seen when, after being extremely hungry, she will suddenly go off her food. This is followed by several hours of restlessness and nesting behavior, along with pacing around, panting, and crying. Labor has begun.

17 "YOU'RE THE LAST, BUT NOT THE LEAST."

Once each kitten is clean, the mother will bite through the umbilical cord, leaving about an inch attached to the kitten's belly. She will eat the placenta of the firstborn kittens, as this will be a rich and readily available source of food at a time when moving away from the nest is unthinkable. She will lick the kittens all over to remove the afterbirth, thus forming the smell bond between kitten and mother (a bond that is often missed when cats give birth through cesarean section at the vet). Then she will rest and wait for the next kitten to emerge. Alexandra has just given birth to her sixth and final kitten, Pearl, and after a cleanup she will join the rest of her brothers and sisters and find her own specially reserved nipple.

NEED SOME HELP, MOM?

Apart from sharing overlapping territories and grooming each other, cats can show great interaction at certain times. When a cat gives birth in a multicat household, other female cats have been known to help out by nipping the umbilical cord, bringing food for the cat, or helping clean the kittens.

18

"WHERE'S MINE?"

Once the litter is born, the kittens are ready to feed. As with most mammals, the first milk is thin but is rich in antibodies to help the kittens stave off diseases in the air they are now breathing. This colostrum will last for several days, after which their normal supply will begin to flow for up to two months. Unlike puppies, who will barge in and feed off any nipple they can get to, a kitten becomes attached to one particular nipple on its mother and will search for that nipple at feeding time. The blind kittens will locate their nipple by smell. If this smell is masked by the mother's contact with some powerful scent, they will struggle to find their rightful place.

19

"KEEP ON FLOWING, MILK, I WANT MORE."

Kittens are pretty helpless when they are born. They weigh about three ounces and are blind and deaf. All they have to go on is their smell, which binds them solidly to their mother and their correct nipple. After three or four days, they will have developed the milk kneading or "treading" technique on their mother's belly, which they use to stimulate milk flow. In later life, cats will re-create this action when they are safe and relaxed in a human lap, or sometimes when just being stroked. At the end of the kittens' first week, their eyes begin to open, revealing vivid blue eyes. Whatever the breed, all cats' eyes are blue at first, changing color after about twelve weeks.

20 "DON'T YOU MOVE A MUSCLE."

After three weeks, when the kittens have developed a degree of movement skills, a feral mother cat will change the location of her nest. She will find a new place and take the kittens there one by one. This she will do by grabbing each kitten by the scruff of its neck in her mouth and dropping it down unceremoniously in its new home. With domestic cats, there is never any real need to do this. It is the maternal instinct borne out of a wild cat's need to give birth in a totally secure, hidden environment and then, when they need to start hunting, take the kittens to a more accessible location. However, domestic cats will still do it, especially if they are being disturbed a lot by overeager children, flash photography, and noise.

TINY TROUBLEMAKERS

Feral or unsocialized kittens who are wary of human contact need to be picked up by the scruff of the neck and then supported by their bottoms. In cat rescue centers in the United States, the more extreme version of kitten restraint is to "burrito" them—i.e., wrap them firmly but gently in a towel so that their claws are not exposed while you begin interaction.

21 "OOPS, THIS FEELS LIKE MOM AGAIN."

The grab by the scruff of the neck looks cruel, but it is actually an effective way of calming a cat down and getting it to stay still. Kittens learn to be transported in this way from one place to another and will freeze their movements so that they don't flail out a paw and hit their mom in the face. This freeze response stays with them, so in later life, when a cat is held this way, it will react as it did as a kitten. It is also an important feature of the mating ritual, when a tom grabs the queen by the scruff of the neck.

THE LIFE AND ADVENTURES OF A CAT

Until the eighteenth century, male cats were known as "rams." Then, in 1760, a book was published in England entitled *The Life and Adventures of a Cat*. The hero of this tale was a male cat named Tom, and from this, all male cats came to be called toms. So if it weren't for this book, we might have grown up watching "Ram and Jerry" cartoons.

22 "WHAT ARE ALL THOSE FLAPPY FLYING THINGS?"

In their first month of life, kittens have to get used to coordinating four legs and moving around. They put on weight quickly, their milk teeth start to come through, and their eyesight improves. Between the first and second months, all the fun starts with the onset of exploration, interaction, and play. The phrase "curiosity killed the cat" is aptly applied to kittens going through this phase, and apart from testing their boundaries and sampling new things with their mouth, they will start to play-fight and play-hunt—stalking and pouncing on leaves in the backyard.

CATS CAN GET SUNBURNED

Cats are good at seeking out sunny spots, but cats with white fur and skin on their ears are especially vulnerable to sunburn. Frequent burning of the ear area can lead to skin cancer, and surgery may be needed to remove all or part of the affected ear.

23 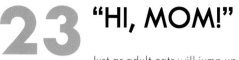 "HI, MOM!"

Just as adult cats will jump up and greet their human parents, kittens will try to do the same with their returning mother, who they can't quite greet nose to nose. The "welcome back" hop onto the back legs is something learned in kittenhood. Not all kittens will have the strength to manage it, and some will try to emulate the hop of their brothers and sisters only to flop hopelessly (and adorably) backward. Balancing on the hind legs is a useful skill for later in life, as so many interesting things happen on raised surfaces that are frustratingly just out of their sight.

LITTLE AND OFTEN

When kittens start to eat solid food, they won't approach meals the same way as an adult cat. Instead of one or two big meals a day, kittens will want to eat a number of small meals. They will arrive at the food bowl, eat a small amount, and then wander off, only to return a little while later. It doesn't mean that the kitten doesn't like what you've given it or it's not hungry; it's a typical eating pattern for a kitten.

24

"OH, PLEASE! JUST A LITTLE SQUIRT. . ."

Kittens can start to eat solid food when roughly a month old. With an ample supply of milk around the house, most domestic kittens will be able to drink milk in a different form, but only as a supplement and not as the main source of nutrition. Some kittens take longer than others to adapt to solid food. However, at two months the mother will stop suckling her litter and they will have to eat solids.

This kitten has just made an unsuccessful attempt to suckle and has been told by its mom in no uncertain terms that she is now off limits! There are specially formulated kitten foods available, and by the age of twelve weeks, kittens should be eating three or four meals a day.

25 "LET'S GO HUNTING."

Mom will gradually introduce her wide-eyed litter to the hunting process in a number of progressive steps. When the kittens are about seven weeks old, instead of eating what she kills on the spot, she will bring the prey back to the nest and eat it in front of them. This way the kittens can learn what to do when presented with a solid lump of food. After that, she may return with a dead animal and "play-kill" it so that the young can see how she dispatches small mammals. Following the lesson on how to kill their dinner, the young will be presented with it to eat themselves. Further lessons involve killing a half-dead creature dragged back to the nest for practice or, more likely, the kittens will accompany their mother on their first hunting trip.

TRICKS OF THE TRADE

Kittens that are not shown how to hunt with their mothers can still learn to hunt themselves, but it is far more of a hit-and-miss affair. Some will use their natural inbred skills and succeed, while others will find hunting very difficult. Kittens that are raised alongside pet hamsters, mice, or pet rats will find it difficult to treat them as prey in later life.

26

"THAT'S ENOUGH PLAYING WITH CHILDREN FOR TODAY."

The key phase for the social development of a kitten is between three and seven weeks, when it is able to start moving around and interacting with its littermates. Before that time, it is important that the mother is allowed to be in charge of her litter and feel safe from constant interruption. Keeping intrusion to the minimum is easier said than done, because kittens develop quickly in this early stage of their lives and everyone wants to see what's going on.

It's at this stage that kittens should start to get some good, positive experiences with humans and begin the socialization phase with their littermates as well as with the people who are going to be their ultimate family.

27 "THIS IS FUN—I COULD DO THIS FOREVER."

As kittens start to play at three weeks old, their sheer lack of limb control means the play lacks coordination. As the weeks go on, though, their movements will become more precise, their vision sharper, their balance more acute, and their intentions clearer. The older they get, the less social play they indulge in and the more they are focused on hunting games. If the litter remains together at twelve weeks, the bids for dominance within the group are clearer to see—there is noticeably more vocalization, and fights can erupt. Even though weaning has long since passed, many experts believe this last phase of learning is very important for a cat's social development.

IT'S A TOUGH DECISION

However important the learning phase between eight and twelve weeks can be for a kitten, the one thing it needs more than anything else is a home. The allure of a cute, cuddly kitten fades, and leaving the sale or adoption of kittens until twelve weeks has to be balanced with the need to place them in loving, responsible homes.

28

"I'M GOING TO SHRED YOUR GINGER EARS!"

As in most confrontations between animals, cats enter fights with a degree of fear. Though they might think they have strength on their side, kitten games have shown them that even the weakest opponent can land blows when cornered. They will use all the tricks in the book to scare their rival and make the fight unnecessary. They will stand on tiptoe on stretched legs to make themselves look as tall as possible. Their fur will rise up to give them a seemingly bigger body mass. They will swivel their ears into an aggressive position and emit a cat "growly yowl." If this isn't enough to make the other cat turn and take flight, they will advance toward it. This tom's threats are enough to bring about total submission from the ginger cat, whose next move will be to bring his back legs up to defend himself from the marauding tom. However, the tom has gotten what he wants already. His forward-pointing ears show that he is not ready to commit to a fight, so he is just threatening the other cat—very effectively.

29

"COME ON, DO YOU WANT SOME?"

Two cats that are hell-bent on fighting—and it's most likely to be toms in a bitter territorial battle—will approach each other almost as though they are in slow motion. The whole time, their eyes will be fixed on their opponent, their ears swiveled into the aggressive position. Should they start fighting, they will flatten their ears so as not to pick up a flailing claw. As they approach, they will raise their heads up and then tilt them over to one side and then the other, sizing up their opponent as they creep forward. The closer they get, the slower their actions become, until they reach a glowering deadlock.

They may still not fight, but if one is to retreat, then it has to do so exceptionally carefully and slowly. Any sign of retreat will be seen as a vulnerability and will be seized on by the other combatant. So the reluctant-to-fight cat will have to reverse very slowly. Here, both cats are swinging a preliminary paw at each other, but in an instant it could become more serious.

30 "IT'S OKAY, IT'S ONLY A SATURDAY NIGHT BARNYARD FIGHT."

Cats will indulge in two levels of fights. There is the "pawing" or "paw-slapping" fight. The claws are out, but neither combatant wants to commit too much to the fight and risk the humiliation of losing. It's a fight at paw's length, a testing-out kind of fight where a victory can be achieved without life-threatening consequences if it all goes horribly wrong. Paw swipes are exchanged in a quick boxing match, and then the participants retreat a short distance to glower at each other.

TERRITORIAL TOMS

Because so many cat fights happen in the dead of night, many cat owners have never seen one. In truth, because the consequences of a cat fight can be lifelong scars, emotional trauma, and more vet bills, cat owners will do their best to stop a confrontation from developing. In the past, a handy bucket of water was the best deterrent against a prowling territorial tom; however, with the advent of high-powered hoses and water pistols, having a "supersoaker" on hand is a good way to keep your cat's virtue (and ears) intact.

31

"LET'S SEE IF YOU'RE AS TOUGH AS YOU THINK!"

The more serious cat fights erupt with a great deal of caterwauling from both combatants. Like sumo wrestlers lunging at each other, they will suddenly launch from a static position and try and knock the other off its feet while trying to bite at the neck; the other will twist to deflect the lunge and attempt to bite back. They will cling on with their front paws while kicking out and scraping at their opponent's fur with their back feet. This frenzy of biting, kicking, clawing, and rolling will last only a few seconds before they separate out and take stock. The bouts will continue until there is a winner or both have had enough and begin the slow, careful retreat.

32

"I WON—I CAN SMELL VICTORY."

After a fight, there is usually a ritual that the victorious cat goes through, sniffing the ground with a big "cat sneer" on its face—i.e., using the vomeronasal organ on the roof of its mouth. Following the small fracas (inset), the aggressive cat on the left clearly senses victory and indulges in a few winning sniffs.

SCAREDY CATS

A fear of cats and kittens is called ailuraphobia. It is said that Julius Caesar, Alexander the Great, and Adolf Hitler were all scared of cats, while Napoléon Bonaparte was scared of kittens.

33 "I'M MUCH BIGGER THAN YOU THINK I AM. NO, REALLY."

Cats will use their aggressive posture when they are challenged by a dog or other frightening animal. Again, they will stiffen their legs and stand on tiptoe to gain maximum height while arching up their back and making their fur stand on end to increase their perceived bulk. Their tails will be high. In addition, they will stand in profile to their potential aggressor to show how bulky they are. This is an instinctive reaction and can be seen in kittens. Should a dog come close, it will be hissed at—closer still and it will be spat at.

OLD CAT TACTICS

Older cats will have learned that it is more likely to be the younger dogs that want to engage in cat baiting. The best form of defense for the cat is to attack, so they will stand their ground, spit, and lash out a paw at the oncoming wet nose rather than give it a target to chase.

34 "I'VE GOTTA GET OUT OF HERE QUICK!"

If a dog is unimpressed by a cat's attempts to scare it off and bounds up in a lively manner, the cat will have to move fast. A cat that decides to run instead of standing its ground is inviting a chase—and dogs love chasing. Whether it's a chase for fun or the re-creation of a chase of prey, it is irresistible to most dogs that are off the leash and outside the command of their owner. The cat may actually instigate a chase by fleeing before the dog has really thought about what it's going to do next. In the event of a chase, the cat will have to use agility rather than raw speed, using its ability to turn quicker, jump, and climb to escape a dog's attentions.

35

"HEY, IT'S CASUAL FRIDAY . . ."

Cats groom themselves for a number of different reasons, but the principal reason is the obvious one—to keep themselves clean. They will spend what seems like hours going through the delicate ritual of removing dirt, grass, and bits of food from every accessible place on their body. (see the cat lick checklist on page 16). One proven way for an owner to tell if a cat is ill is to look at the condition of the cat's fur. Fur out of place or remaining matted and dirty signals that the cat is too weak or distracted to attend to this most important of cat functions.

36

"THIS'LL TAKE AGES."

Another reason for cats to groom is to maintain their own personal smell. Often it is possible to pick up a cat and smell the strong perfume or cologne of its owner in the cat's fur. If this smells strong to you, imagine what it smells like to the cat! Once a cat is released from a human embrace, it will start licking itself all over to restore its essential cat smell, its identity. It will also lick to sample your smell and enjoy the experience of that—providing it isn't a barrage of cheap deodorant.

A WATERPROOF COAT

The action of licking helps keep a cat's coat waterproofed. Just as a human's sebaceous glands are stimulated by constant washing, so the glands at the base of a cat's hair respond to the tugging action of grooming. Cats can groom this way because they have rough, abrasive tongues covered by numerous barbs, or papillae. Located at the tongue's center, the papillae are backward-facing hooks that are effectively the "teeth" of a cat's "comb."

37

"YOU KEEP MISSING THIS SPOT—USE YOUR PAW."

Cats can be upset when they are covered with strong-smelling substances, so a caring owner should be wary of dousing herself with eau de toilette and then giving the cat a big hug. Typically, when a mother gets her kittens back from a brief handling experience by the humans in the house, she will want to imprint her smell back on the kitten and make the kitten her own again. In this instance, the mother is licking the spots that the kitten finds difficult to lick itself, particularly the top of the head. In time, after watching Mom do it, they'll get the hang of it.

38 "GOTTA KEEP THE INSULATION IN PLACE."

Licking at ruffled fur isn't simply a cat's fastidiousness, a need to be clean, tidy, and presentable at all times. In cold weather, it will lose a lot of heat through its fur—and the more a cat's fur is ruffled, the more heat it will lose. This is not so important to cats in places like California, but for Norwegian forest cats it is essential. Cats need to operate in temperatures from 25°F to 100°F, so the maintenance of fur is critical to their well-being.

39

"I NEED TO COOL DOWN."

Apart from an uncanny knack to find the coolest places, licking is the cat's main method of keeping cool in summer. Like dogs, cats do not have very many or very big sweat glands, so they cannot lose temperature the way humans do. Instead, they will lick big swaths of their fur, depositing as much saliva on it as they can. The evaporation of the saliva takes heat away from the cat's body and cools it down. While the human body supplies its own coolant from skin pores, the cat has to deposit its coolant on the surface to achieve the same result.

40

"CAN'T YOU SEE I'M BUSY?"

Cats sometimes lick as a nervous reaction. If they find themselves caught in a situation they can't immediately escape from, or are stuck in a tense encounter with other cats, they may start to lick. It's a sign that the cat is uncomfortable or agitated, and one way of relieving the tension is to start to lick. Cat psychologists call this "displacement grooming." It's grooming for the sake of having something to do.

CLEANING AN OBSESSION

The almost-obsessive licking of fur is rarely a problem in short-haired cats, but displacement grooming in long-haired cats can be a problem when fur balls form in their stomachs. These are usually vomited out before they get too large, but it is something that owners of Persian cats have to be aware of and can combat with regular brushing and combing.

41 "I ADMIT IT, YOU'RE MY FAVORITE."

"Allogrooming" begins at birth for a kitten. It is the act of being groomed by another cat, and in a close-knit cat family this may go on all their adult lives. "Autogrooming" is another term for self-grooming and will go on whether a cat has a close family or is a complete loner. Autogrooming will begin at around three weeks old.

Allogrooming is less about hygiene and more about the cats maintaining a strong bond than anything else, though another cat is able to get to places that are difficult to reach, such as behind the ears. This motherly lick is an example of bonding, rather than necessary cleaning, as it's an area the kitten could easily reach itself.

A FAIR OLD LICK

Cats can lose more water from their body through licking than they can through urinating. It's thought to be one of the reasons why cats mark with scents less than dogs. They just don't have the volume available.

"GET A LOAD OF THIS."

Tomcats like to mark out their territory by spraying the ground with urine. Just like their canine counterparts, they make sure they leave their mark against upright, vertical surfaces such as fences, bushes, and trees. The scent can then be fully appreciated at nose height by every passing feline. In fact, the scent of a tom is so strong that humans notice it too and, along with the occasional torn ear and the wandering, it is one of the hazards of owning a sexually active tom. However, females and neutered toms will also mark territory as part of their daily routine, but since their output is less pungent, it is much less noticeable.

NINE LIVES DIVIDED BY FIFTEEN

The average cat can be expected to live for about fifteen years. It is believed that having a cat neutered will increase its life span by two or three years. As most cats are neutered, the average life span for an unneutered cat is twelve to thirteen years.

43

"WHO'S BEEN MARKING ON MY SEAT?"

Cat spraying is the feline calling card. Like flyers that are posted on street corners, the cat goes on its rounds reading all the interesting scenting posts in its territory. A lot of information can be gleaned from the freshness and composition of the scents—they will tell a cat exactly who has passed by, how recently, and whether or not they are sexually active. When they finish reading, they will add their own comment at the end—large if they have a full bladder, small if they don't have much liquid to spare. Indeed, cats will continue the marking motions even when they have run out of liquid, so strong is the habit. In this picture, the cat is eager to sniff out the previous occupant of the chair, a cat that was sitting with its rear directly on the seat and which has inadvertently marked the chair with an interesting scent.

"THIS IS MY BACKYARD, AND I'M RELAXED."

Since house cats don't have to rely on their hunting activities to feed themselves and their offspring, they don't have to defend the food supply within their territory. So whereas wild cats have huge territories, urban cats have very small ones. What's more, the fences and boundaries we put up around our backyards make it very easy for cats to divide up areas among themselves. Territories will still overlap, because cats rarely meet when they're out and about reading the "daily news," but if they do, they're more likely to avoid each other than start a fight. Generally speaking, a male cat's territory will be about ten times the size of a female cat's territory, allowing him to monitor the sexual state of several queens.

HAVING A CAT FIGHT

Cats have had some bad PR in the past—"having a cat fight" is a phrase used to describe a noisy clash over some petty disagreement. In reality, cats are much more tolerant of others coming into their territory than incredibly fussy dogs who flare up over the slightest challenge to their status.

45

"THIS IS A GREAT CHILL-OUT ZONE."

Where several territories overlap, cats can develop a kind of neutral zone where they can meet without conflict—or with very little conflict. You could call it a cat "drop-in center." Though cats are by nature solitary creatures, they can drop in and drop out of contact with other cats, especially in areas that have no defined ownership.

A typical cat congregation point on the Greek islands is the harbor, where cats all gather to see if they'll be thrown scraps from the returning fishing boats. In this photo, the ginger cat is quite happy to hang around in the presence of the black cat, although even in periods of light sleep a cat's ears will provide constant surveillance.

46 "I'M IN. I'M OUT. I'M IN. I'M OUT AGAIN."

Whereas dogs will pick up scent messages on their walk maybe once in the day or sometimes twice, cats are more diligent at the task. For the homeowner without a cat door, the cat will be constantly in and out and meowing at the door. It is not because they can't make up their mind. They like to go out and check their territory—maybe do a little extra marking—smell the scents, and then return to the security zone that is their house. Because scents degrade over time and the cat has the freedom to travel without their pseudo parent (unlike a dog), they like to go out and get the news while it's fresh.

SQUEEZILY DOES IT

Cats can squeeze through the narrowest of spaces, aided by their whiskers to judge the distance and their flexible spines—cats have thirty vertebrae, double the number of human vertebrae—and also due to their lack of a collarbone. It means that any space they can squeeze their head through, they know they will be able to get their body through, too.

47

"I'M FEELING RELAXED."

The ears and tail of a cat give the strongest clues as to how it's feeling. With the ears, five different messages can be conveyed: relaxed, alert, under stress, defensive, and aggressive.

The relaxed cat's ears point forward and to the sides. They are flatter than usual. They are monitoring the situation over a 360-degree field, but they don't expect trouble anytime soon. They're on standby.

WATCHDOGS OR WATCHCATS?

Cats have better hearing than dogs and much better hearing than humans. Cats can hear sounds up to 65 khz, while humans hear only up to 20 khz. What's more, they can rotate each ear 180 degrees to get a precise fix on where a sound is coming from and can react to sound far quicker than dogs.

48 "I'M FEELING INTERESTED."

A cat's eyes will be wide open when it moves from a relaxed "standby" position to "alert." It has heard an interesting noise, one that has to be checked out. The ears are erect and pointing forward in the direction of the noise. Sometimes the ears will be diverted by another sharp, interesting noise and one ear will swivel toward the direction of that noise, giving the cat double coverage. These are the ears of a cat that is either anticipating trouble and straining to identify a strange sound, or a cat who has heard a fascinating rustle and is about to switch into hunting mode.

EARS SOME FACTS ON CATS
- Cats are thought to respond better to women because women have higher-pitched voices than men.
- Kittens are born with closed ear canals, which begin to open nine days after they are born.
- A cat has thirty-two muscles in each ear.

49

"I'M UNDER PRESSURE."

Just as humans under stress develop nervous twitches, so a cat that is stressed will develop twitchy ears. At first sight, it might look like your cat has got some kind of ear infection, but it is also a sign that it is agitated and—as in the case of a wagging tail (see page 112)—suffering from conflicting emotions.

THE ORIGIN OF THE SPECIES

The domestic cat is thought to be a descendant of the African wild cat (Felis silvestris libyca), which is also known as the Egyptian or Kaffir cat. Evidence of the cat's existence alongside humans dates from the Egyptian Middle Kingdom period of around 2000 BC. They first appear in tomb paintings, and some were even mummified.

50

"WHOA, I'M SC-A-A-A-A-A-RED!"

A frightened cat will flatten its ears down against its head. This is a gesture that is normally the prelude to flight, with the cat reacting in fear and then making as fast an exit as possible. Tucked-in ears are not only aerodynamically more streamlined, they are less susceptible to bites and swipes from the claws of attacking cats. In this position, the hearing of a cat is reduced, so they don't like to maintain the ear position for too long. The cat on the left is very anxious about the swaggering threats of the aggressive cat on the right and is about to take flight.

51

"WANT TO TRY IT, THEN?"

Aggressive cats have a very particular ear position. They are swiveled so that a portion of the backs of the ears are visible from a (hopefully intimidated) cat gazing at them from the front. In profile, they are halfway between the "interested, alert" height and the "scaredy cat" flattened position. It is intended to strike fear into a more docile cat by communicating that it is ready to fight at a moment's notice. In reality, it is also ready to retreat should the other cat surprise it and accept the challenge.

These are the same two cats that are sizing each other up on page 108, and from this angle we see the full threatening pose of the aggressive cat: the turned ears, the staring eyes, and the body held high. Trouble is brewing.

52 "I'M NOT SURE ABOUT THIS."

A wagging tail from a cat is a signal that it can't figure out what to do next. The sign is one of conflicting emotions; it wants to take one course of action, but it's got some serious reservations. The wagging cat tail is like a massive thought balloon from a comic strip—it says "I haven't made my mind up yet." Cats often wag their tails on the doorstep when they want to be let out at night but aren't sure if it's such a good move. When a cat makes up its mind, the tail stops wagging. In this photo, the cat wants to get closer to a bird on the lawn, but from experience it knows that its color is a dead giveaway the moment it emerges from the long grass. So, what to do next...?

ANGER WAGGING

Cats will also wag their tails in anger situations as a sign of frustration. They will weigh the consequences of an aggressive move. The calculating cat wants to do two things at once: it wants to exert its dominance or expel another cat from its territory, but at the same time it's fearful of what might happen if things go wrong.

53 "GIVE IT A REST, I'M NOT IN THE MOOD."

The lashing tail is a sign of great agitation and emotional turmoil, and the degree of upset is reflected in how much the tail gets lashed from side to side. The signal is intended to be a warning to stay clear, don't come near. This long-haired Tonkinese (right) is trying to give this signal to the cat on the left. It has fluffed up its tail and is lashing it from side to side, and has also assumed a submissive position. Although it is crouching, its ears reveal that it is not particularly frightened by the aggressive move of the other cat.

54 "SNIFF IF YOU LIKE."

The vertical tail is a sign of open friendliness. It is a sign of approachability. Although human owners never take them up on the offer, a cat is saying "Hello, I'm going to sniff you. Do you want to sniff me?" In cat etiquette, it is a social convention.

55 "I'M NOT IN A SOCIAL MOOD. GO AWAY."

The wrapped-around tail is similar to when a human crosses his or her arms in front of the body. Cats are using their tails—whether consciously or subconsciously—to put a barrier and some distance between themselves and another cat. It's a defensive posture that means they don't want to interact with others.

56 "LOOK, I'M TRYING TO HAVE A CATNAP HERE."

The flicking tail is a sign that a cat isn't fully settled and there is some small, irritating detail that is getting underneath its fur. Though it's not serious enough to get the cat to move, it is a sign that the irritant is still around.

57 "WE'RE REALLY CLOSE-KNIT FOLK."

Tail twining is a friendly gesture. Kittens will try to twine their tail around their mother as a request for food. Adult cats can also be seen entwining their tails. This is another form of scent rubbing, a way to transfer scent from the anal gland that has been spread onto the tail. However, it usually comes as a conclusion to the normal friendly greeting of rubbing heads or sniffing noses and rubbing cheeks first.

THE CAT FANCY

The first cat show was held at the Crystal Palace in London in 1871. Although the United States had hosted many small events before, it was in May 1895 that the first national American cat show was held at Madison Square Garden in New York. The Best in Show award went to a female brown tabby Maine cat named Cosie.

58 "I DON'T THINK IT CAN SEE ME . . ."

Domestic cats are pouncers, not chasers. Unlike their big-cat cousins, who will indulge in long pursuits of young wildebeest on the plains of Africa, the domestic cat likes to creep up and pounce. First they need to stalk their prey, and in this their approaches are very similar to that of their larger relatives. Using scent and sound, they will locate their prey, and keeping as low a profile as possible, they will stealthily approach. However, a cat's conflict between the need to move forward and yet be as soundless as possible will start its tail wagging (similar to the cat on page 112).

At night, when it is hunting rodents that it cannot see, the wagging tail is no hindrance at all. On an open lawn, when it is stalking a bird, the wagging tail acts like a flag and draws attention to it, scaring the birds off. This is why cats are much better nighttime hunters of small rodents than they are daytime hunters of small birds. Even so, many species of wild birds have been endangered by cat kills, and the American Bird Conservancy launched its "Cats Indoors!" campaign in 1997.

59

"PRACTICE, PRACTICE."

Cats can often be seen flicking objects into the air in order to chase them or jump up at them and simulate an attack. Kittens will do it with toy balls, and older cats also employ the same technique, sliding a paw underneath the object to give it a quick flick up or backward. With this action, cats can practice their fishing techniques.

Cats have different attacks for birds, small mammals, and fish, and will sit at the side of ponds or streams ready to employ their fish-flicking technique. If a fish comes into range, they will use the quick-flick paw to land the unwary fish. Sometimes cats will practice with prey they have already caught or, in the case of kittens, with small objects like balls of wool. In fact, it comes so naturally to kittens that they will do it even without hunting guidance from an adult.

60 "IT'S LUCKY I'M BEHIND THIS WINDOW."

For those owners who've seen it happen, it's one of the most amusing things to watch. Your cat is sitting at a window, carefully observing its backyard domain. All of a sudden, a bird lands just outside the window, no more than a few feet away, but safe from the pouncer behind the glass. Your cat is transfixed and may adopt a crouching stance, even though it cannot spring out to catch its prey. As the bird flies off, the infuriated cat bares its teeth and works its mouth in a silent chattering motion. The cat is re-creating its "killing bite," the biting action it exerts on its prey once it is captured. It is an instinctive reaction, and it will do the same thing outside when a bird narrowly escapes its clutches.

CATS' EYES

Though they will use their acute hearing and sensitive whiskers for hunting, an important part of their equipment is their night vision. Cats are believed to have eyesight that is six times better than a human's, though their field of view is very similar—around 185 degrees

61 "YIKES! I THOUGHT YOU WERE DEAD."

Cats target their killing bite at the neck of small mammals. Their aim is to kill or incapacitate their prey by using their strong canine teeth to sever the spinal cord. Even if it is not killed, the bite will usually prevent the victim from escaping. As some owners have seen from cats' mouthing at birds tantalizingly beyond their reach, this involves a rapid, repeat motion of the jaws. Bites aimed elsewhere might not stop the mouse in its tracks, and some small mammals have learned to hunch their bodies so that the cat cannot figure out where to bite. Inexperienced young cats may make what they think is the fatal bite, relax to admire their kill, and be surprised when their prey gets up and runs off.

JUDGING THE DISTANCE

A cat sways its head from side to side in order to judge the distance between itself and prey. The closer an object, the more its position will shift each time the cat sways its head. After the long creep into position, this action is the indication that the cat is about to pounce. Rodents move quickly and cats attack from a short distance, so the attack needs to be very precise.

62

"I DON'T WANT TO GET TOO CLOSE."

Young cats or inexperienced cats may paw a small rodent around through fear, rather than exercising their hunting instinct. Cats will chase and kill rats, but they don't enjoy the size advantage that they have over mice, shrews, and voles. Some of the urban rats that have grown large from eating discarded food are a force to be reckoned with. Their teeth can inflict a serious wound on a cat, so instead of going in close for the killing bite, cats try to beat them up first. A cat will use its arm's length advantage and two sets of sharp claws to batter the rat and severely wound it before moving in for the kill. When they do finally approach to inflict the fatal bite, they are wary. This wariness can be seen in young cats who aren't convinced that a vole they have battered is actually dead. In these instances, the cat isn't reenacting the thrill of the hunt, it is just making extra sure that its victim is dead before sticking its face up close.

63 "WOW, THIS IS FUN. AND I CAN MAKE IT LAST ALL AFTERNOON."

Cats often play with their prey rather than giving it the instant dispatch of a killing bite. Many owners can't bear the thought of their pets being so cruel, but it is a natural reaction to their pampering. In the twenty-first century, small rodents are rarely a problem around the house.

They get into attics and other places where cats don't have regular access, but on the whole the biggest rodent problem is the rat, and cats don't play much with them. Because a cat doesn't have to provide its own food and rarely gets the chance to hunt, when it does finally capture a "small moving snack," it's excited. It doesn't need the "food," but it does enjoy the experience of exercising one of its natural instincts. To prolong the thrill of the hunt, it will catch and release, catch and release, using retracted claws and soft bites until it is tired of the game. Even a yellow-and-green mouse that doesn't run away can provide the cat with hours of fun.

64 "LOOK WHAT I'VE BROUGHT FOR YOU."

One of the hazards of being a pet owner is receiving gifts from your cat in the form of half-chewed or half-dead animals deposited outside the door. The cat has noticed that its "parent" is not so good at going out and catching its own prey, so the cat provides the kill as a service. When a cat is teaching its kittens to hunt, it will bring back similar offerings for its litter. For this reason, the neutered female or the littlerless female will bring back more animals than a neutered male, as it is performing an instinctive role.

JUST WORKING UP AN APPETITE

For the "large kitten," which is basically what all domestic cats are, there is no need for the prey to be eaten to satisfy hunger. When a cat wanders off after killing its prey, owners assume they have killed for the sport and nothing else. This isn't the case, because most cats will move away from their prey and loiter nearby before returning to the scene. In effect, the cat is taking a breather before returning to eat the carcass. Small mammals that have been playing dead use the opportunity to escape, and some cats may return to their meal only to find it gone.

"IT'S MY SEAT AND I'M THE TOP CAT. DEAL WITH IT."

A group of cats living in the same space will determine their own hierarchy. There will always be a top cat. Generally speaking, they will coexist happily side by side, especially where they share the same food source. Within this group, some may be intolerant of others that are close to them, while some form close social attachments. The degree of harmony will stem from a great number of factors, including the maturity of the cats, their breeds, the spaces they have to share, and, most importantly, the socialization they enjoyed when they were kittens.

THE CAT GOD

The ancient Egyptians are known to have worshipped the cat goddess, Bast (sometimes known as Bastet). She was said to be the protectress of the royal house and the Two Lands of Egypt. Her role was later expanded, and she became the protectress of women and the goddess of joy. In Egyptian society, cats were revered and killing a cat was a crime punishable by death.

"I'VE GOT THE BEST VIEW."

A cat that is forever trying to reach a high place in your house is likely to be doing more than just trying to get a good view. It's strange to think of cats in the same light as military campaigners, but in a multicat household, to be top cat you need to dominate the high ground.

Cats will try to dominate any place that is up high: the owner's bed, windowsills with great views of the backyard, kitchen counters, and furniture placed near doors. Of course, they can't cover all these bases at once, so the interesting situation for owners to observe is when the top cat arrives at one of his or her favorite locations only to find another cat occupying the spot.

MUMMIFIED CATS

In ancient Egypt, great respect was paid to cats. When the house cat died, the family would go into mourning, shaving off their eyebrows as a mark of sadness and respect. Cats were often mummified and even given jewels, trinkets, and little mummified mice to take into the afterlife with them. Dedicated cat cemeteries have even been found in parts of Egypt.

67

"LOOK AT MY BIG, MACHO WALK."

A top cat adopts the swagger of a feline that is undisputably the leader of the gang. It will walk on tiptoe but with a lowered head. Its whiskers will be angled forward, and its ears will be alert but angled sideways. This posture is meant to tell other cats to keep their distance, or, if they do want to approach, they will have to do so in a subordinate way. A cat that walks directly toward this cat while giving it a direct stare will be asking for a fight.

68 "I'M JUST TINY AND HARMLESS."

To show a top cat that it is no threat and that it would prefer peaceful co-existence, a subordinate cat will crouch down low to the ground while presenting its profile. It will avoid direct stares if it can, preferring nervous glances. Its ear will be pressed right down against its head, and it may be preparing to flee at any second. Its tail will be tucked around its body, both for its protection and to present as small an object as possible to the top cat's eyes.

TOP 20 MALE CAT NAMES

1. Max	8. Smokey	15. Rocky
2. Sam/Sammy	9. Toby	16. Sooty
3. Simba	10. Spike	17. Bailey
4. Charlie	11. Rusty	18. Felix
5. Oliver	12. Jake	19. Sebastian
6. Oscar	13. Buster	20. George
7. Tiger	14. Sylvester	

69 "I'LL GO FIRST, YOU CAN WAIT."

Apart from the key spots within the shared domain, there are other pointers as to who is the top cat. We all know that there is one "warmest spot in the house" during the winter months. Who gets to be in it says a lot about the hierarchy the cats have established. Top cats will deem that certain areas are their prerogative—around the food bowl, for instance—or they might make a big fuss about being the first to use the litter box after it has been cleaned.

70

"I'LL JUST KEEP RIGHT OUT OF EVERYBODY'S WAY."

At the opposite end of the cat social order is the low-status cat: the youngest, most timid, least aggressive cat in the group. While the top cat will take center stage in the household, the low-status cat slinks to the edges. In a typical cat community, it is the youngest and least socially mature cat that takes the position on the bottom rung of the status ladder.

When a new arrival believes that its status should be further up the scale, huge conflicts can arise. That is why new cat introductions to a multicat household can be fraught with problems.

The low-status cat will be one of the last to eat if feeding positions are limited. It walks with its body low to the ground and rarely engages the other cats in direct eye contact.

71 "I'VE GOT MY EYE ON YOU."

Top cats will do a lot of posturing to maintain their position as leader of the gang. As with all animals, they will try to outbluff a rival before choosing the potentially debilitating option of a fight. The way they exercise their control over other cats is by direct staring, by rubbing up against them, and by holding their bodies high in a threatening, I'm-bigger-than-you gesture, as we have seen throughout this book. The games of kittenhood are played out all over again.

Another tactic to keep the other cats in line is play-mounting. Neutered or intact cats, both male and female, will use mounting as a way to dominate another cat, the same way they did before they were sexually mature.

"I KNOW MY PLACE."

Many factors influence a cat's place in the hierarchy: size, age, health, fearfulness, sexual and social maturity, family links, the age at which they were removed from their mother, and the size of their territory.

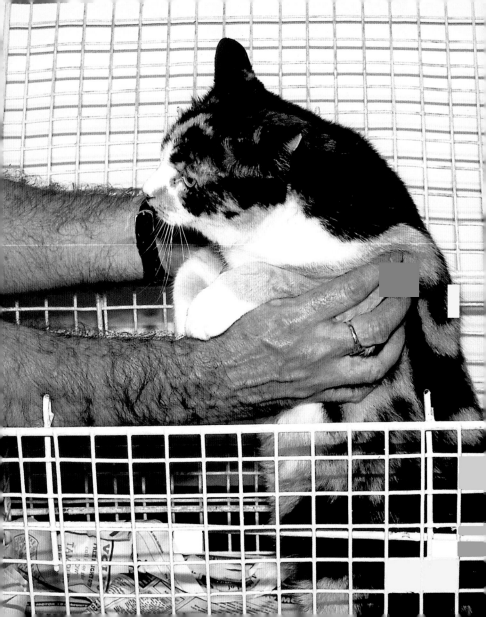

72 "I DON'T THINK I'M GOING TO LIKE THIS."

Because so many of the things that influence a cat's position within a hierarchy are changeable—health, age, social maturity, etc.—their position in the hierarchy can change as well. If a cat is ill and feeble, their status will drop considerably. The pecking order is a fluid thing. Cats will be agitated by the introduction of a new member. The way in which owners introduce a new cat to a multicat household is absolutely vital to the harmony of the feline occupants. Get it wrong and there can be long-lasting consequences (plus a few unpleasant smells around the house). Cats seem to enjoy being part of a hierarchy and knowing their position.

A CAT'S TALE

There are many bones in a cat's body, but around 10 percent of them are in the tail alone. The domestic cat is the only species that has the ability to hold its tail in the vertical position while walking.

72 "I'M NOT GOING TO BLINK FIRST."

Cats that give you an unblinking stare are trying to make a point. The stare is one of the key elements in aggressive encounters between both cats and dogs and is used to threaten or dominate. Staring hard at a cat will make it feel uncomfortable. If it stares back, it is trying to make you feel uncomfortable.

The opposite is the slow blink, sometimes known as the "cat kiss." Slow blinks indicate that a cat is relaxed and friendly.

TOP 20 FEMALE CAT NAMES

1. Molly	8. Sophie	15. Angel
2. Misty	9. Bella	16. Beyonce
3. Princess	10. Shadow	17. Sadie
4. Samantha	11. Callie	18. Katie
5. Lucy	12. Daisy	19. Jasmine
6. Missy	13. Cleo	20. Peaches
7. Kitty	14. Chloe	

74

"FIGHT, FIGHT, FIGHT!"

The belly-up posture of a cat that rolls onto its back is a defensive-aggressive action.

By rolling onto its back, it cannot actively engage its rival because it cannot move anywhere. However, by rolling over, it has all four sets of claws ready to engage, as well as its mouth ready to bite. A cat springing on top of it will have to be careful not to sustain at least one or two vicious swipes from its rival and must judge whether the fight is worth it. The cat that rolls over is giving the aggressor top-cat status, because it clearly believes it might fight, but it is also showing that it is ready to fight if necessary. It's a warning, but a "losing" warning.

The posture, as can be seen from this picture, is one learned at a very young age. The kitten on its back has rolled over and is about to employ all its weaponry on the kitten that has decided to attack.

"NO, I DON'T WANT YOU TO TICKLE MY TUMMY."

If a strange human comes into a household where a dominant top cat rules the roost, the top cat may challenge the newcomer. A cat will use the scent glands on the top of its head to rub against the legs of an unfamiliar visitor and then sit back and try and stare out the visitor. This isn't a welcome—it's a challenge. Humans often mistake the rubbing head as a sign of welcome and acceptance. When they try to pick the cat up or stroke it—as a reward for this marvelous feline welcome—they can end up with a scratch for their troubles.

A more catlike greeting is to hold out a hand or a finger to be sniffed. If the cat likes what it smells, it might run its jaw along your hand to say, "Okay, you're cool." Once your scent has been committed to the cat's memory, you can start getting acquainted.

76 "NOPE, THERE'S NO ROOM UP HERE."

Cats view territory in a house like a chessboard with several different levels—the higher, the better. This is why cats are so eager to walk along the old-fashioned dado ledges seen in period homes, occasionally knocking down vases and pictures as they go. These ledges are the perfect walkway for a cat, who is able to look down on its domain. Fashion catwalks aren't so named because of the temperamental behavior of the models backstage! In the average room, there are many levels—floor, windowsills, armchairs, low tables. There are also the doorways, the room's center and edges, and safe places under low furniture. How cats arrange themselves in a room in a multicat household tells you a lot about their status and ranking.

THE NOT-SO-LAID-BACK CAT

A cat's pulse can vary between 140 and 240 beats per minute, far quicker than a human's pulse. Whereas a cat can average around 190 beats, a human's average is 72.

"I'M THE KING OF ALL I SURVEY."

If cats aren't allowed on the furniture in a multicat household, this can cause the cat to become upset. One of the ways in which cats exert their influence is by controlling the higher areas of the house. If they are all banned from climbing the furniture—a kind of cat communism, where every comrade kitty is treated the same—they cannot demonstrate who is the top cat. This, in turn, might lead to the cats demonstrating their superiority in a more confrontational way, and squabbling can break out.

Also, if homeowners suddenly purchase new furniture, the carefully mapped-out territory of the multicat house has to be mapped out all over again. This is why cat trees—a structure like the scratching post on page 156 but with more levels—are useful for giving the cats the vertical territory they need. It's also a good way for you to see who's the top cat and who's the bottom cat.

78

"THIS SHOULD PUT HER IN THE MOOD."

It has been quite a battle for the tom to be first in line, but now that he is, he has to make the most of it. The mating act will only last a few seconds, and he has to make sure that his queen is perfectly in position. To get her into the position where she can accept him—known as lordosis—he starts a strange act in which he paddles his back legs on either side of her rump. Eventually she will move to the correct position and they can begin.

79

"STAY RIGHT THERE— THIS WILL ONLY TAKE A FEW SECONDS."

Tomcats may be the bullies of the backyard as they attempt to be first in line in the mating queue, but they are quite timid when it comes to dealing with females. It is the female who lives up to her name and queens it during their courtship. The tom is just a humble (and anxious to please) courtier. In the mating act, it is the tom who fears being battered by the aggressive and dominant female. When the two animals mate, the tom will grab the female by the scruff of her neck in his jaws—not as an act of domination but as an act of self-protection. He is counting on what she obeyed so instinctively as a kitten—being picked up by her mother—to keep her still. The moment he withdraws, she will be clawing angrily at him.

80 "SO, HOW WAS IT FOR YOU?"

The cat mating act is very brief—but then comes the tricky part. A tom's penis is covered in short spines that point away from the tip. These allow for pain-free insertion, but the result is a very painful retraction. When the male withdraws, the female suffers a great deal of pain as the spines abrade against the walls of her vagina. The photo seen here captures this moment of intense pain that the female cat suffers at the end of the mating act. However, there is a good biological reason for the pain. Cats do not ovulate until after they have been mated, and this first painful shock triggers the start of the ovulation process.

81

"UH-OH, I'M GETTING OUTTA HERE!"

In the short term, toms don't want to stick around to face
the consequences. They have a very good reason for
making a fast exit—the female will lash out and give the
male a very hard time after putting her through such pain.
But the drive to mate is so strong in the female that after
half an hour or so, she has forgotten how bad it all was and
is ready for another mating act. In this photo, we can see
the tom's fear by the position of his ears as he hightails it
(literally) to safety.

82

"I'VE GOT THE CAT'S WHISKERS."

A set of fully functioning whiskers gives a cat great awareness, especially in the dark. Depending on the breed, cats will have around twenty-four whiskers, twelve on each cheek arranged in four rows.

Cats are sensitive to air flowing over their whiskers, and the whiskers can be used to detect small air currents passing over objects. When cats move in for the kill in the dark, an escaping rodent is located when it brushes against the whiskers. In effect, the whiskers are like a radar system, used for moving around in the dark, avoiding objects, and homing in on nocturnal prey.

83 "I CAN SEE YOU!"

Because humans regard sight as their key sense, the fear of going blind is immense. However, cats can communicate through smell, touch, sound, and sight. What's more, they are nocturnal creatures who navigate and maneuver with great skill in very low light, using their memories of shapes and spaces as well as their whiskers and vibrissae to sense nearby objects.

If a cat becomes blind, it can still lead a perfectly fulfilling life. Owners can help the cat by keeping furniture movements within the home to a minimum, but the cat will take it all in its stride.

It is important to note that an owner should make sure that a blind cat's whiskers are kept in very good condition and are not chewed by other cats in the household.

THE NICTITATING MEMBRANE

Cats have an extra eyelid, known as a nictitating membrane or haw. They usually only appear when the animal is ill or extremely dehydrated and a trip to the vet is needed. They are often more visible in Siamese, Burmese, and Tonkinese cats.

84

"YOUR SHIRT IS WAY TOO BRIGHT!"

Cats' eyes are extremely sensitive to light because they possess a light-reflecting layer known as the tapetum lucidum, which helps capture more of the light entering the eye. This aids them tremendously in their role as a nocturnal hunter.

However, because their eyes are so sensitive, the great intensity of direct sunlight needs to be controlled. That is why cats have pupils shaped like slits. Cats adjust the level of light entering their eyes by adjusting the size of the slit and by lowering their eyelids. A cat that lowers its eyelids in strong light isn't necessarily sleepy; it's just trying to avoid being blinded.

85 "I'M NOT AS QUICK AS I USED TO BE, SONNY."

Old cats lose their range of movement, their joints stiffen, and they become less flexibile. Whereas at one time a cat could leap onto a counter, in old age it will struggle even to get onto a chair. An owner of an older cat needs to be aware of the cat's increasing frailty and help it out, whether by lifting it up to high places or by helping groom areas of fur that it cannot reach. A cat will usually slow down only in the last one or two years of its life. This is not long in a life span that may last ten to fifteen years.

TEACHING OLD CATS NEW TRICKS

Old cats are very much like old humans. They hate new things and adapt to new situations poorly. Buying a kitten "to put some life into the old cat" is not a good move, as the old cat will tend to get irritated and resentful, while the kitten can't understand why the old cat won't play. Moving to a new home can also be stressful for an old cat.

86

"WOW, THAT SMELLS INTERESTING."

Cat owners will notice that their pet occasionally pulls a kind of grimace or sneer when it comes across a strange or fascinating smell. The reaction isn't because a cat is offended by what it smells; in fact, it's so excited it wants to get more. It's known as the flehmen response, the cat opens its mouth and draws back its upper lip while drawing in a breath to be sampled. Like a connoisseur of fine wine, it considers the smell for a while as though mulling over the bouquet of a great vintage. The cat is using the vomeronasal organ in the roof of its mouth, which is sensitive to airborne smells. Part of its ability to read so much detail into the scent markings of other cats comes from using this taste-smell organ. In this photo, the tom is using the cat sneer to check out the scent of the female near him. Sadly for him, she is not in heat, even though she is being friendly.

87 "WOW, OUTTA SIGHT, MAN!"

If you've seen a cat rolling over a plant in your garden, licking it, biting it, chewing it, rubbing up against it, and generally making an exhibition of itself, that plant will be catnip (Nepeta cataria). It is a strange sight to see: the most reserved of cats suddenly behaving like it's on a drug-enhanced road to ecstasy. After ten minutes, the effects wear off and the cat will return to normal (though it will remember precisely where the catnip is located). Roughly half the cat population is immune to the charms of Nepeta cataria, but a good percentage will fall under its spell.

IT'S A QUESTION OF TASTE

Kittens will not go near catnip. However, at three months old a change occurs. Those that enjoy the sensation of the plant will start to seek it out, while the teetotalers will continue to avoid it.

88 "I'M KEEPING AN EYE ON THINGS."

Cats may sometimes look like they're daydreaming because they are not focused on one thing in particular, but they are actually quite attentive. This is due to their excellent peripheral vision. They will only stare at something intently if they are getting a fix on a moving target, getting ready to pounce. Monitoring a situation without making it look like hard work is quite easy for them to do.

When a cat is being stared at—whether by a human or a cat—it will stop and check out the level of the threat before continuing on its way.

"IT'S RIGHT IN FRONT OF YOU."

Scientists disagree about exactly what colors and shades a cat can see, but it's believed they can see blue and green, though red is still open to debate. One area they cannot see is right in front of their noses. Cats have a blind spot directly under their nose—which is why, if you put some tasty morsel down for them to pick up, they can have a difficult time seeing it.

89

"I'M STILL LISTENING."

Few animals and even fewer pets sleep as much as the domestic house cat. They can spend between twelve and sixteen hours curled up asleep on a sofa or dozing in the backyard (bats and opossums sleep longer, but they don't make for very good pets). This is because cats are obligate carnivores with a protein-rich diet. They get their calories in large amounts, unlike herbivores, who must graze all day to maintain their body mass.

Thus, cats can live the life of leisure—the extended kittenhood. About 80 percent of a cat's sleep is light sleep, and given that this cat has fallen asleep on an open stretch of grass, it is likely that the sleep is shallow, with its ears maintaining surveillance the whole time.

90

"GOT TO GET THESE OLD CLAWS OFF."

Cats claw at furniture for a number of different reasons. The first is to pull away the old, blunted claw sheaths and reveal the new, sharp claw sheaths underneath. A cat will use up a set of claw sheaths and then move to the set underneath, like a craft knife with several extra blades in the handle.

Pulling them off is difficult, hence the need for the scratching post—or, more accurately, a "clawing post." Scratching posts do not sharpen up claws like a file or a sanding pad; rather, they help cats remove their old claws.

FAT CATS

It has been estimated that 50 percent of pet cats are overfed and overweight. Of these, around 20 percent are obese. One of the reasons, apart from the obvious fact that owners supply them with too much food, is that a lot are kept indoors and simply do not live the lifestyle of an active carnivore.

91 "THIS IS MY STRETCH-AND-SCRATCH FITNESS WORKOUT."

Apart from getting rid of the old claw sheaths, cats scratch to exercise the extension and retraction of the claws from the feet. An average house cat's day includes little use of its claws in hunting or climbing, so a scratching post is the equivalent of a cat gym, where it works on that part of its physique. Because the front paws also include scent glands, the action of clawing helps spread a cat's scent over the area. It's not only using the post as a gym, it is saying very clearly, "This is my gym." This kitten has found the best place of all to exercise its claws.

OBLIGATE CARNIVORES

A cat has thirty teeth: twelve incisors, ten premolars, four canines, and four molars. They are designed for eating raw meat. While dogs can be omnivorous, cats are carnivores and turning them into vegetarians can lead to protein and vitamin deficiencies.

92

"I WANT MY DINNER AND I WANT IT **NOW!**"

Cats can be very vocal when they're hungry, and it's quite easy for a loving cat owner to respond to their pet by opening the food can right away.

It's best to feed a cat at regular times. Feeding on demand is likely to reinforce the behavior that a cat has to ask for its food and pester you for it. By responding to a cat's request for food, the owner is making it more likely that the pestering will happen again.

"IS THIS A CAT BOWL?"

A cat will prefer a shallow and wide food bowl to a deep one because it will constantly bump its sensitive whiskers on the sides of a deeper, narrower dish.

93

"I'M THE INDISPUTABLE LEADER OF THE CAN."

A feral cat is one that was once domesticated but has now reverted to the wild. Feral cats tend to live in colonies in large, open spaces such as abandoned factories or around farms— places where they can find just enough food by hunting or scavenging and yet won't be disturbed.

The important behavioral distinction between a feral cat and a house cat is that the feral cat will be unsocialized with humans and will not want to be picked up. However, getting close enough in the first place might be difficult—they are wary of human contact. Feral colonies rarely overlap suburban territories, not because of the lack of open spaces but because of the immense problems that can result when the territories overlap.

94

"IT'S LIKE BEING A KITTEN ALL OVER AGAIN."

The extended kittenhood of house cats shows up most obviously in an action known as "milk treading." When a kitten is feeding, it will knead away at the mother cat's belly, pushing with one paw and then the other in a gentle, rhythmical motion. At the same time, it will emit a characteristic satisfied purr. A cat that jumps into your lap and re-creates this same kittenlike rhythmic pawing is going back to the first six weeks of its life. It enjoys playing the part of the kitten—when life was uncomplicated, when it felt safe and loved. The fact that it might also be accompanied by drooling and chewing of clothes isn't so great for the mother substitute, but it's the total affirmation that you're seen as the cat's present mom. This cat is milk treading while it is being stroked.

95

"CAN I HAVE A LITTLE PRIVACY, PLEASE?"

Although cats are strongly territorial, they bury their feces because they prefer not to waft their odors around the neighborhood and cause conflict. While active tomcats will see it as a way to advertise their control of an area, burying is the act of a cat who doesn't want trouble. That is why cats will go to great lengths to find a suitable location that can be dug over after the business is done. This doesn't hide the odor altogether because, as many gardeners know, cat mess is pungent stuff. However, burying does reduce the odor, so the smell wafting from backyard to backyard doesn't pose as strong—or as serious—a threat as it would if it were left on the surface.

Cats will also want to cover their feces in the litter box at home, and they regard it as a private occasion. For this reason, placing the litter box in a quiet corner is much better than leaving it in the middle of a room.

96

"I LOVE A GOOD CHEW IN THE MORNING."

Cats love chewing grass in the backyard, but why they do it is still the subject of debate. Some think that they do it to induce vomiting—through their constant grooming, cats end up with fur balls in their stomachs and need to get rid of them.

As cats rarely seem to do much more than crush the grass between their teeth, an alternative explanation is that cats use it to improve their diet by chewing folic acid out of the blades. Folic acid keeps cats from becoming anemic.

BIG CATS, SMALL CATS

The heaviest house cat recorded was a mixed-breed Australian cat named Himmy, who tipped the scales at a massive 45 lb. 10 oz. One of the smallest ever was a Himalayan-Persian named Tinker Toy, who was only 1 lb. 8 oz. when it reached adulthood.

97

"I LIKE YOU AS A FRIEND, NOT JUST BECAUSE YOU'RE A FOX."

If cats are socialized with other animals when they are kittens, they are less likely to try to hunt them later in life. For peace and harmony in a multipet household, it is wise to introduce your kitten to all kinds of playmates before it starts to regard them as meals on legs—or, in the case of dogs, something to be feared.

This kitten and fox cub play together just like littermates, though the fox's rate of growth means that it will soon be too big for rough-and-tumble games.

98

"I'M A REAL SOCIAL ANIMAL."

Alhough cats have their social moments when they engage with other cats, they can take company or leave it. Siamese cats are not built this way. A Siamese cat loves company, preferably another Siamese, but a child, dog, or adult human will do. What's more, they are good at adapting to routines and will happily wake up when you do and sleep when you sleep. Their reputation as cruel creatures—as portrayed in Disney's animated film *The Lady and the Tramp*—is not at all true. Siamese cats make great pets. They are the cat breed that behaves most like a dog.

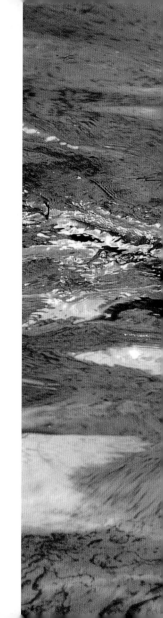

99

"YES, I KNOW I'M DOING THE DOGGY PADDLE."

Unless it's in a bowl, cats have a great aversion to water. But not the Turkish swimming cat, the Van. They are called Van cats because they were discovered in the 1950s near Lake Van in eastern Turkey.

Vans are big, active cats with white bodies, colored heads, and very expressive, flouncing tails. Their most amazing characteristic, though, is that they will fetch things from the water, and they enjoy swimming. It is probably the intense summer heat in their Turkish homeland that has produced this love of water. A Van will jump into lakes, streams, ponds, baths, and even sinks, but owners should make sure it doesn't get too much chlorine by going into swimming pools.

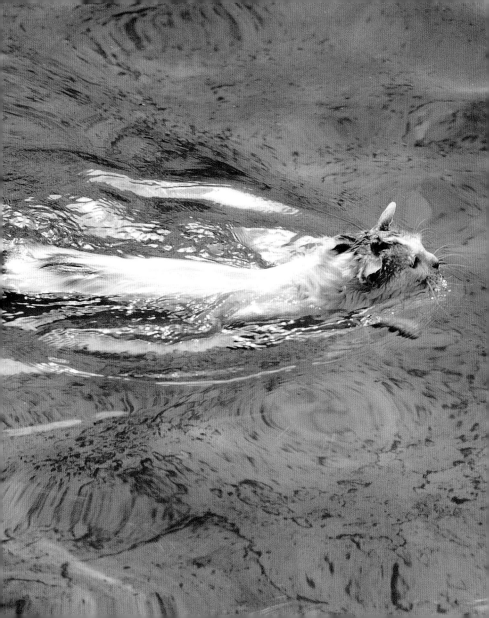

100

"WE ARE BURMESE, IF YOU PLEASE."

The Burmese is the closest to "cat Velcro" you can get. Burmese cats, like Siamese cats, need a lot of human contact—perhaps a Burmese needs even more than a Siamese—and will follow their owners from room to room in close attendance. They will want to sleep in the owner's bed, hop onto the owner's lap whenever he or she sits down, and carry on life as though they were a conjoined twin. They are the ultimate companion cat, actively seeking out contact with their owners. They do well as indoor or apartment cats.

If only to add to their human characteristics, they also have a very stubborn streak.

INDEX

ACKNOWLEDGMENTS

The publishers would like to thank Warren Photographic for supplying the bulk of images for the book.

© Warren Photographic/Jane Burton: Pages 13, 17, 21, 26, 29, 30, 33, 34, 37, 41, 43, 44, 47, 50, 54, 57, 58, 61, 65, 67, 71, 72, 73, 75, 76, 78, 81, 82, 84, 86, 89, 90, 92, 95, 101, 108, 110, 114, 117, 119, 120, 126, 129, 130, 134, 137, 138, 141, 142, 145, 147, 151, 152, 155, 156, 161, 163, 164, 167, 170, 174, 177, 178, 189, 193, 194, 197, 198, 200, 203, and 205.

© Warren Photographic: Page 133.
© Warren Photographic/Kim Taylor: Pages 49 and 68.
© Warren Photographic/Mark Taylor: Page 190.
© Getty Images: Cover image; pages 98 and 159.
© Anova Image Library: Pages 7, 10, 14, 19, 23, 25, 38, 53, 62, 97, 103, 104, 107, 113, 115, 116, 122, 125, 148, 169, 173, 181, 182, 185, and 186.